Gaussian Curves

From A to Z

 GP

By Bishnu Goswami

Gaussian Curves
From A to Z

By Bishnu Goswami

 GP

This page intentionally kept blank.

Dedicated to Rini Chandra.

<u>Preface</u>

The world is a very fascinating place. Emerson rightly said "When I first open my eyes upon the morning meadows and look out upon the beautiful world, I thank God I am alive." Unfortunately, as on the time of this writing, the beautiful world is reeling from a global pandemic. We will emerge victorious in the end, that is for sure, but we should also spare a minute for the frontline workers and thousands of people who put forward a valiant fight. Let us all learn from this crisis, and better fight against future ones, as we gradually extinguish the last capsids of these viruses.

Knowing beforehand that certain random phenomena follow a specific distribution, we can model the real world with incomplete data. Data is always, in a way, incomplete and if it was not so, the world would become a boring place. Therefore in this book, with some data we have, I tried to map them in some normal (Gaussian) distributions. Hope you will enjoy it!

Thanks a ton!

-Bishnu Goswami
15th July, 2020

Introduction

There are five (basic, for more precise uses, there are many more) sense organs and they all collect data. Their organization, analysis and presentation, combined, when combined is called statistics. One of the most important successes of statistics lies in predicting the outcomes of random events, events which have outcomes that are not predetermined. This prediction, in the applied world, is often the sculpting block which is chiseled away into a fine statue with more real-world data.

One of such widely used sculpting block is the 'particular distribution', also called the bell or Gaussian curve. This creates a graph which models binomial distributions (such as the outcomes of a coin toss) when the number of 'turns' approaches infinity. Although very important in the theoretical foundations of statistics (especially when considered with the central limit theorem), it is widely used in our day-to-day life, albeit in less mathematically precise way. From snap judgments about on whether to trust an astrological prediction (only for believers in astrology), to corporate boardrooms decisions to hire a candidate based upon standardized tests, the curve and the position of the point sought is often used.

These Gaussian curves look like a hill and have their "hilliness" quantified in according to the statistical variance and the mean of the underlying set of data, technically with two main assumptions which prohibits infinite variance and mean. Of course, in many cases these means depend on subjectivity, as in the case of astrological predictions, and in other cases they are very objective (when normed diligently with quantitative data).

In this book we delve into the interesting manifestations of Gaussian curves in various arenas of our day-to-day lives. This book is not for budding statisticians looking for formal treatments or number crunching, however. This book rather aims to show how the observations are distributed along both the quantitative and qualitative aspects of our life, how they deviate, and what that means for the rest of us. A touch of humor is also in the cards, according to our in-house readers!

Some of the topics are discussed relatively simply, for the relatively younger readers and some go into a little more detail. In the experience of the author, this makes reading more interesting and can often motivate the readers for further exploration on the related topics!

THE MAIN PART

Aliens

Aliens in this context refer to extraterrestrial life. Extraterrestrial lives are a hypothetical. Nobody knows if there is any, but *many* theories and hypotheses point to the fact that life should exist on other planets rather than our own. This was perhaps a show of humility by the humankind. We are, after all, perhaps nothing *special*!

The Drake equation, sometimes popularized in laymen science books, or sometimes in some popular science TV shows, tries to model the chance of the *existence* of aliens who can *communicate* with us. It is expressed as the product of:

- Mean rate of star formation
- Fraction of stars that have planets
- Mean number of planets that could support life per star with planets
- Fraction of life supporting planets that develop life

- Fraction of planets with life where they develop intelligence
- Fraction of intelligent civilizations that develop communication
- Mean length of time the last cohort can communicate for

All these terms are multiplied to get an estimate of the number of alien civilizations which can communicate with us. There are a number of important observations to be made here. First, the abundance of the term 'mean' means that *means* are very important in statistics. When there is a mean, there is a normal distribution lurking somewhat close, or what is called the Gaussian curve itself.

Secondly, it shows that how deceiving equations can be, if someone is not careful. In school days, we sometimes take that equations and theories to be *true*, as a matter of plain 'binary' fact. But most of them, in a deeper level of analysis, are actually clouds which can rain only sometimes, depending on a slew of additional, very important factors. The reliability of equations can also range from very reliable to <insert your least favorite politician here>-esque. The last few terms in the multiplication, from the Drake equation, to give an example, are so difficult to estimate that the final result can vary by several orders of magnitudes (It can be 10x, 100x, 1/10x or $1/100^{th}$, between different estimates).

Thirdly, the equation and its surrounding context also tell us that how difficult it is to quantify things that aren't backed up by truckloads of useful data. But there is also hope in the horizon in the form of the Gaussian curve, with which we can model random variables with unknown distribution. After that is made, we chisel away the clunky bits and create a bust out of the gently sloping concrete of the curve.

Now, let us get down to the final construction, *aliens*. Contrary to the school of thought in the first paragraph, the great majority of us perhaps love ourselves a bit too much. And there's nothing wrong with that. If we loved ourselves too less, mirror companies would go bust!

What facts point to the possibility of our self love? It is in the depiction of aliens. Most aliens have a humanoid form. To make them distinct and stand out, but not too much, there are some popular themes to the humanoid extraterrestrial. They are often diverged and mixed with the reptiles, therefore when they might have eyes, their external ear (pinna) is not there. Again, with perhaps a hint of narcissism, the aliens are sleek and slender, often even more than us humans. We are after all above animals and their brawny pectorals and biceps to compensate for their lack of intelligence!

Some aliens, in shows depicted for kids, have shorter and chubbier construction, in sync with neotenous traits which adults love (as they are like their children) and children can see as being peers with. This trait should also go in popular category.

Less popular ones, but very scientifically sound alien-form hypotheses, vary. Humanoid forms are very unlikely to be a common alien form, if we go by scientific principles and data which tends to repeat itself, at least here on Earth. The progress of life is best modeled by natural selection, although there are very important specifics we need to be aware of (such as it is not the best way to model humans in modern times). By this train of thought, extraterrestrial lives have a much higher probability of being simple unicellular or acellular organisms.

On the other side, we have constructions which seem neither *that* popular nor scientific. Unemotional, faceless electronic aliens, for example, are difficult to emotionally connect to and are unscientific in the sense that their circuitry are far too rudimentary to support themselves unless we stretch our imaginations very much!

So there we have it!

Popularity

Electronic
robots
without
faces.

Huamnoid,
bipedal, and
lizardy!
For kids:
Fluiffier, cuter
and more
friendly!

 Possible
Reality

Simpler
organisms,
unicellular
or acellular.

Amorality

Morality is deviously tricky. Barring a few fundamentals, they vary in space, time and perhaps in other hidden dimensions if they exist. There is so much variety in moralism, moral policing and moral sciences that it is trying to tackle all of Google's page requests in a homemade server. A good alternative here to stay on the page and the alphabet is to consider amorality. Amorality is not what is *not* moral. Amorality is not immorality, the latter being actions that are believed to be wrong and/or improper.

Sapience is one of the criteria for considering things to be perfectly amoral, which is not an oxymoron. A brick is free from moral transgressions, or is indifferent to so, whether it falls on someone's head by gravity and storm, or if it is thrown by someone against somebody else. But what might be technically more correct can possibly differ from the view of nonspecialists. They might consider perfectly amoral non-sapients to be outside the equation. Therefore, with the majority view, the criteria of sapience cannot be the central bulk of the Gaussian curve here.

'Do as you would be done by' is a slightly terse verse of the Bible (Matt. 7:12) but is well respected across cultures worldwide, at least in theory if you take a quick sampling through a questionnaire. Neglecting this, especially through a cheerful indifference, is something many people would consider to be amoral. Corporations, which have the sole motive to increase profit for the shareholders, can be regarded in the same vein, and with large popularity, across wide spheres of societies. Of course, more community oriented societies, and societies having a thrust-upon or organically grown seeds of socialism, might disagree on this. Many people in these societies can also see businesses as plain evil and money to be cause of many of the society's problems. There is a rift between amorality and immorality here, but they can possibly be lumped together as besides the precise definition of the two terms, the lack of morality in itself is deemed immorality by many people. These topics of discussion should occupy the bulk of the curve-enclosed area.

Amorality has some extremes when religions get involved. In the Bhagvat Gita, one of the holy scriptures of the Hindus, it is said that Lord Krishna defines lack of morality in a case-by-case basis. In this decision tree, the morality varies by the social rank, age of the person and the large scale relationships among larger families. Hinduism, just by an example and with no intention to single this particular religion out, also puts no absolute restriction on the morality of killing. This is sharply contrasted by cultures who believe on the universality of morality and egalitarianism; and proponents of the latter sometimes find the other view very difficult to internalize. Hume also pointed out that many of the 'greatest crimes' were allegedly compatible with superstitious piety and devotion. Here the complexities between religion and amorality are twisted like a complex knot.

However, with the rising trends of universalism by globalization, the discussions in the last paragraph are getting marginalized on the curve. It would, therefore, be away from the center.

Popular-criteria to be considered

Evidences, and theoretical justifications

Do as you are done by, AKA, the Golden Rule, basic care towards society

Case by case basis, depending on age, social rank, clan etc

Most of the greatest crimes are compatible with superstious piety

Charlatans

While some nationalists and/or local-cultural fanatics (it is very important to note that they might occur independent to each other) of developing and newly industrialized countries might disagree, the history of the old developed countries and even the USA has many things to offer to the countries that have a catching-up to do. One of them is in the context of charlatans. Their history well documents the wicked saga(now this is indeed an oxymoron) of charlatans.

The very definition of the world charlatan derives from the French word with the same spelling. Charlatans were sellers of medicine who used to advertise their products with music and a street stage show. The word can also be traced back to a village in Italy, Cerreto di Spoleto, which was known for its abundance of quacks. Either way, the charlatans seem to be an ongoing problem throughout the world, and perhaps even rising in places which were previously not exposed to the outside world.

Charlatans come in many shapes and sizes and traits of nefariousness, but they still can be modeled using the golden Gaussian curve. Some of the traits of charlatans come up very commonly and if we are aware of those, we can spot them easily (and also help others who are less experienced spot them) and subsequently thwart the plans of evil.

One of the traits of charlatans that make them easy to spot is their stress on haste. Sometimes the object of import in some works of fiction, charlatans makes us take major (either per person or recouping cost through number of clients) decisions in a very short amount of time. This is coupled with other tactics to make us yield to the charlatans plot. This stress on haste can manifest in many forms, such as giving discounts on the spot which will not be valid later, or stressing that he has only one bottle of magic medicine left for sale. Like in dating, a stubborn stress on haste is one warning sign.

There are also traits of charlatans that we can protect ourselves from just by knowing a few things. One is that get-rich-quick schemes are a hoax. Second is that pyramid schemes are by their construction loss-inducing. Third is that there are no Nigerian prince or princesses.

However, very well-read, educated people still fall into these traps every now and then. The author has seen university postgraduates peddling multiple pyramid-schemes through a lack of understanding. Many seniors in developed countries or even in newly industrialized ones still fall prey to Nigerian princes, even though they might not be from Nigeria (and obviously not princes!).

These are the central bulk in our topic of discussion.

For specific age groups, especially younger women, the related charlatans are often catfishers or other scammers. They create a fake online persona, such as a young member of the Military in active duty and ask for financial or other help. Older women, especially those widowed or divorced, are also sometimes the victim of these evil charlatans. A basic competence in fact checking is often enough to deter these, and numerous websites exists to inform us how to exactly verify tall claims pertaining to a specific profession in a specific country. They are not 100% foolproof, but the are very effective to catch the 99%.

Stolen valor is another example for countries who take a pride in their military, such as the USA. These charlatans dress up as military, perhaps due to a combination of self-delusion and the military discounts they get in many civilian establishments.

On the ends of the curve, we have charlatans who have also fooled themselves akin to the way they fool others. Sometimes these are harmless activities which waste few things other than time. Some form of extreme ritualism in some religion's derivative, or political systems falls under this category. Not all cannot be branded in a timeless fashion though, as we can take the example of libertarianism . This school of thought, both in their right and left backed ideology, was sometimes labeled as works o charlatans and sometimes they have got a greater acceptance in the society throughout history.

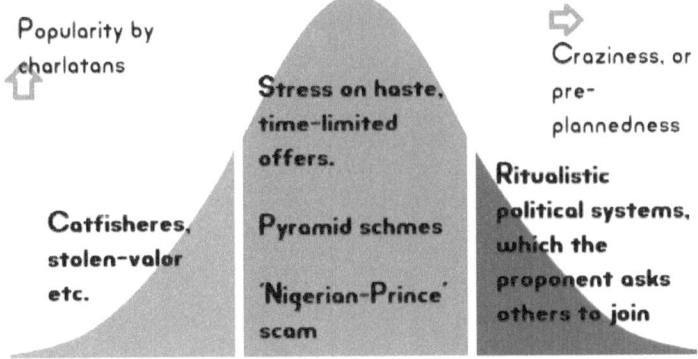

<u>Dreams</u>

Dreams can be of the things what we feel while we are asleep, or can be the things that we feel about when we are awake, or in the process of a metaphorical awakening.

The literal dreams are a very active area of research with few useful findings, especially on the fundamental questions such as "Why do we dream?". Therefore, here we discuss the dreams that we feel when we are awake.

Ask school going children what their dream is. Many of them would want to be a teacher, a doctor, a member of the armed forces (depending on the country) and perhaps an engineer. These dreams are often force-fed, and perhaps not completely unjustly so, as the recipients are perhaps not mature enough in their understanding of the intricacies and scopes of every jobs. Many children, however, do follow these dreams and make them a reality. These dreams should be included in the central block.

When we shift our attention to adults, especially younger adults, their dreams are sometimes nothing to write home about, if we were expecting something extraordinary. Many prefer, if they are honest, money, a beautiful wife, vacations to exotic places, and perhaps electronic gadgets or motor vehicles. The definition of ordinary is its huge number of elements in the set. Ordinary is what that is commonplace. Very satisfied with our categorization, we should put these themes of dreams in the central category.

On the ends of the Gaussian curve, we can find examples of the extraordinary. These examples can be political, as they have a very long reach.

"I have a dream" was a speech by Martin Luther King. That dream was in no way ordinary, that was a dream which was backed by extraordinary courage, dedication and the subsequent follow through. That was an extraordinary dream of a more equal state and civil liberties. Before King, Gandhi did have a similar dream for India, and examples are littered throughout history. But they were not at all ordinary. They were refreshingly, breathtakingly rare. These should occupy one end of the curve, where they were nothing if not noble.

The 20th century had many groundbreaking advances for the humankind, such as going to the fairyland of the moon, but it was also plagued by darkness which was never seen before. Nazism, Fascism and even some aspects of socialism (which the modern world takes a squinted-eye look at sometimes) were dreams that turned out to be evil for the great majority of the people These can be, despite their minor resurgences every now and then, can be lumped into one end of the curve, perhaps with the label of evilness attached to them. They can be a very big chapter by themselves if we like to pursue.

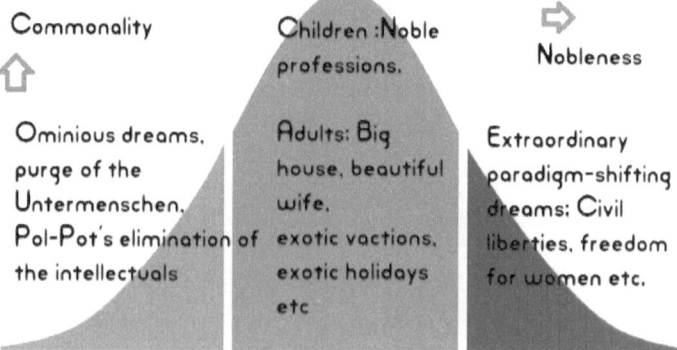

Commonality

Children :Noble professions.

Nobleness

Ominious dreams, purge of the Untermenschen, Pol-Pot's elimination of the intellectuals

Adults: Big house, beautiful wife, exotic vactions, exotic holidays etc

Extraordinary paradigm-shifting dreams: Civil liberties, freedom for women etc.

Education

If we take formal education, as opposed to education in a more holistic sense, the equation nicely fits the bell curve. That is the same as the Gaussian curve, it's useful to roll our tongues differently sometimes.

Education is strongly correlated with literacy. The literacy rate in all developed countries was almost 100% since a long time. For many former and current developing countries, it is steadily growing towards the magic three digit number. Therefore, contrary to our first declaration in the previous paragraph, education do not nicely (nothing real-life ever fits *exactly*) fit into the same curve independent of time.

In the conventional use of the term, formal education has a specific purpose. In most places of the world (frowning at you, uncle Sam), education is partly or wholly subsidized by the government. They do have to justify to the tax-givers. Therefore we do see strongly selective admission to the lucrative medical, engineering and reputed management schools. Seats are limited as the grade goes up, with Ph.Ds, especially with funding, having a much more limited number of seats. What this results in is that almost everyone can have a chance to have high-school education and perhaps a satisfying college experience.

This is useful as not everyone needs to be doctors of philosophy. There is a lot of work still required by the standards of our technology which is to be done by bare hands. There is lot of people management to be done where PhD's might not be the best way to train oneself (maybe except in noncriminal psychology)! And people-management jobs in the higher tiers have much greater salaries than the longbeards with Ph.Ds(to offence to them, I'm doing one myself)!

The curve here, with our trimmings, is so simple that we need not further elaborate. Here it is.

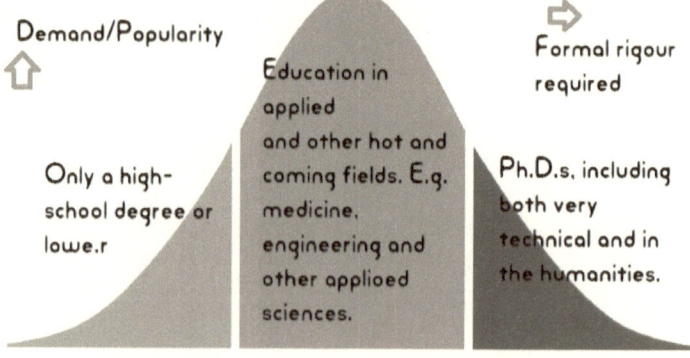

Demand/Popularity ⬆

Formal rigour required ⮕

Only a high-school degree or lowe.r

Education in applied and other hot and coming fields. E.q. medicine, engineering and other applioed sciences.

Ph.D.s, including both very technical and in the humanities.

Entrepreneurship

Liberal economies in the former developing countries, after the 70s and in some places after the 90s, paved the way for entrepreneurship to flourish in previously unseen numbers and growth percentages. It is defined as the readiness, ability and follow-through of a, usually, business enterprise in order to make a profit. In very simple words, it is the set up of a new business venture, with a tinge of innovation.

The majority of entrepreneurs are of the garden variety small businesses. They are often easy to model and their financial transactions rarely stray from a few orders of magnitudes. They usually have a well defined scope and proven small markets. Therefore, they do not usually go bust. As the comfort of relative predictability wraps most small businesses, droves of people try to cash on in them, thereby increasing direct competition. Many of us are familiar with small service companies having competition being propped up next door, in the next year if not the month. This partially explains why they do not scale up very easily. This type of entrepreneurship forms the bulk of the curve.

When talking about small business, it is natural that large corporations will come up. They are excluded from this comparison as they can be regarded as a separate topic of discussion on corporate entrepreneurship, which can complicate the discussion a bit more than on the scope here. This is done all the time for many statistical analyses, sometimes to push a specific extreme viewpoint. Readers should always be aware of this seemingly minor but very important subtlety. Or they should stop watching news.

Another form of entrepreneurship is the popular line of start-ups. They usually start from scratch and they often have (sometimes they pretend) a scalable business model. These entities crave for growth and are hungry for capital. They can also be called greedy in some collectivist contexts. They are not as common as small businesses, which can cover almost everything in less conglomerated parts of the globe. These start-ups should cover one end of the curve, with the second axis of central-aggregation of money.

Another type of entrepreneurship which is quickly rising in many parts of the world is the concept of social entrepreneurship. In developed countries with more of a laissez-faire economic system, the monopolies are mostly growing faster than the smaller businesses, thereby increasing the economic divide, irrespective of the fact of the overall economic development. Here social entrepreneurship, which excludes a profit-first motivation, enters the picture. These types of entrepreneurship aims to help the weaker economic sections of the society to develop themselves through innovation and profit-sharing among the masses. This category should occupy another end of the curve, with the second axis showing the value of dispersion of capital.

Hence, we can have the following picture:

Popularity

Variety of small
businessess with
proven slow-
growth models.

⇨

Less-central
aggregartion of
money

Start-ups with
a high growth
or quick-death/
takeover.

Service-oriented
labor intensive
companies.

Social, more
inclusive
entrepreneurship.
Profit-motive is
absent/minor.

Fiction

Fiction refers to stories that are not primarily based on the real world and the commonly accepted standards which can call a supposed fact a fact. Outside of courtrooms, where fictions are a lie and is riddled with criminal intent, fiction is what we understand as novels, novellas and short stories.

Based on imaginations, how can fiction be quantified in a curve? Isn't fiction like the vast swatches of clouds behind the peaking mountain-scenery of the human imagination? Those perhaps cannot be quantified, as every landscape photographers can tell you. The scenery changes perpetually as the water flows in the river. No two times are the same.

But when we use kite as a metaphor to fiction , this can perhaps be understood easily. Kites roam the sky and cover vast spaces in three dimensions. Yet at the end of the day, they come down. Kites can also be brought down using hooks and other such assortments. Fiction can be quantified if we break it down, though sometimes unjustifiably leaving some quantities out of it. This does not necessarily refrain the quantification from being a possible work of art. Mona Lisa is beautiful only up to some specific levels of magnification, and in some particular wavelengths of incident light.

One such metric is to consider the sales. TV channels show us, soap-operas sell. Sports sell. They have a large market. Classical music is god's own language to purists but they are few in number. In the topic of fiction, we also have a similar curve. Nobel-prizing winning works of literature gets the nod of fine wine by a *very* select few. It takes prep-work to understand the depth in those works of fiction.

Fiction is commonly categorized into two broad categories. Genre fiction refers to the works that sell and are very popular among the masses. These, however, do not commonly represent the pinnacle of literary prowess. That is represented by literary fiction, which sells less but is usually more nuanced, introspective and has the depth in their characters. Literary fiction writers often work under a patronage, such as an university. Therefore, genre fiction should represent the central bulk and literary fiction should represent one of the ends. On the second axis, as we can see, we can add the term quality. This quality can be quantified in the complexity of characters, such as the avoidance of common plots and stereotyping.

A third category of fiction can be those work which work on very simple plots, and have a surplus of hormone-inducing spices added. This is sometimes made for the first-time readers and young people of a specific age and maturity bracket. With the innovation of cheap internet, however, their popularity might be on a decline. Texts can only go so far!

Works like the one involving various shades of gray is borderline between genre fiction and the aforesaid third category. Young women (and men) tend to love (lust after?) them, in a higher percentage than the author would like. This might make the curve skew to the left, but for simplicity we avoided that here.

Now we have a workable curve for fiction!

Popularity

TV Soaps, sports fantasy games, music in fictional themes mostly of the pop genre.

Quality, quantified by plot-complexity etc

Cheap, crass type of fiction, once staple of worker class's freetime.

Literary fiction: including those winning Nobel prizes and Booker prizes.

Intelligence-Quotient

Intelligence is a very multi faceted term. It is often defined by a general capability of the mind to reason, plan, think abstractly, solve problems and to learn quickly. Sometimes, there are additional factors included in the equation, such as the supposed "emotional intelligence" and also aspects of both short-term and long-term memory.

One longstanding method to measure intelligence has been the utilization of intelligence quotient(IQ). It is a score that denotes 'intelligence' using standardized tests which usually requires no specific advanced background knowledge. Some tests include a section of testing the abilities in one's native language, while others do not, the latter being called a 'culture-fair' test.

The methodology to measure intelligence, and even defining the meaning of the term itself, is not without its fair share of controversies. The discussion here is quite nuanced, so we won't delve into it that much. But the major criticisms include the subjugation of one section of the population by another. Sometimes, contrary to the intuition, some jobs have also been denied to people *for specifically* having a higher IQ than the mean. For this particular example, one applicant for the position of a policeman was denied the job because his high intelligence would find all-day patrolling boring!

Despite its share of controversies, it is a matter of fact that IQ is extensively used in educational and employment opportunities. Sometimes it is labeled specifically as such, and at other times it is not. Tests such as the SAT, GRE and entrance exams to many prestigious management schools often share a large set of questions which are common in IQ tests.

The IQ scores are modeled along a rather quintessential Gaussian curve. Therefore it offers an intuitive example to study the practical uses of the curve itself.

Consider two dices which are rolled. What will be the likely sum of their up-sides? Unbeknownst to many people who lose money in casual carnival gambling games, the number 7 has a much larger chance of cropping up, at slightly less than 17%. The sum of 2 an 12, however, will only come up with a chance of less than 3%. This can be approximated by a Gaussian like curve, signifying the binomial distribution of probability. Perhaps only the very high IQ persons (especially in the mathematical sections) can intuitively infer this at the first glance. The rest of us have to do a study and refresh of probability.

The distribution of IQ, and the standard Gaussian curve, is very steep near the center, and gets more and more flatter, nonlinearly, as we go near the ends. Two standard deviations later, there is less than 2% of the population. In IQ, the standard deviation is around 15 points. Two standard deviations later, the IQ of 130 and above, are present in only around 2% of the human population. This should remain relatively unchanged each time the population is normed. Therefore, persons with IQ above 160 are very rare in the population, and tests often fail to measure anything above that with reliability and the very important drawback of limited sample size.

Failure to understand this dispersion can result in rather nonsensical hypothetical questions such as "What happens when someone has an IQ of 250?"

With the concepts of binomial distribution, sample size, norming, linearity and standard deviations being introduced (and many indirect terms related to them not explicitly written about) , it might be useful for the reader to look up these terms in more detail. These are very important terms in real life for both small and large matters of importance. The bare basics can be gleaned from many sources, but at the very least rudimentary mathematical expressions should also be included, and not just text as in this book.

And here is the perfect curve for this topic.

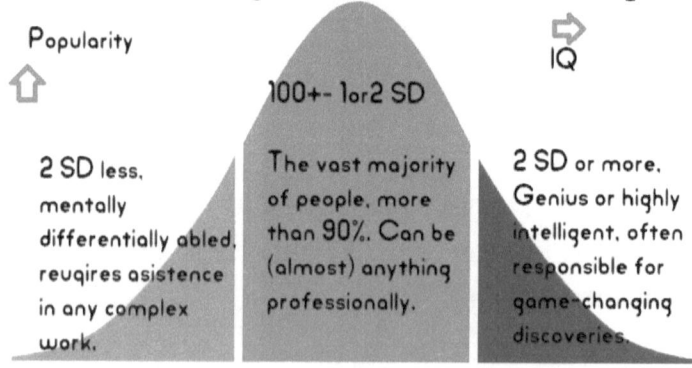

Popularity

⇧

⇨
IQ

100+- 1or2 SD

2 SD less, mentally differentially abled, reuqires asistence in any complex work.

The vast majority of people, more than 90%. Can be (almost) anything professionally.

2 SD or more, Genius or highly intelligent, often responsible for game-changing discoveries.

Kayfabe

Kayfabe is the collective term used to denote events in a show which are scripted, but nevertheless is portrayed as reality. Successful kayfabe might not always be logical, as it calls for some pausing of critical thinking. But that's usually not a problem for the kayfabe or the show, as most of the entertainment industry relies on some variety of such omissions. Man is not a rational animal, despite it might be a having a greater predisposition to, in Aristotelian times!

A prime position to examine on kayfabe is the business of show wrestling, where some very globally popular companies (or *the one* company) are involved. While tackling the commoner themes in one axis, we have to think the variable for the second axis.

The commoner themes include the heel and face turns. For the uninitiated (these are much widely known and circulated than basic math/science topics, much to our dismay!), heels are the evil guys while the faces are anything but. Perpetual goodness and perpetual evil becomes boring after a while, so one character stays a heel for some time and then changes back to face after a few months or the next year. This advances the scripted storyline of the show, and keeps the fans entertained.

While a dominant theme in kayfabe, these turns are not always very successful. Sometimes characters have a dominant gimmick or some personality traits that make them very fit for one role, and not the other. This caused some major problems when one wrestler (who represented a kayfabe empire like the Greek) was pushed into a forced face role. This was, as widely supposed, done for merchandise sale reasons. However, it backfired when the audience booed so loudly that the audience microphones had to be (allegedly) turned down!

However, there are often ways to circumvent that first-impression stereotype. A heel can be made face without making him a nice guy by making him stand up to corporate-management oppression. It might not be the best example for impressionable young minds, but it works very well in show business.

Another popular theme is character origin stereotyping. It might be an antithesis to equity and globalized egalitarianism, but it kind of sells. Sells very well actually!

The third popular theme is the thematic motifs from day-to-day culture specific friendly conflicts. These include schoolyard bullying themes (which is rapidly losing popularity due to obvious reasons), rap battles and rivalries involving sports jerseys and other related themes.

All these three put together can represent a significant portion of the central bulk of the curve.

While inquiring about the second axis, we can take an egalitarianism-hierarchialism political axis. This is relevant because many of these show business employees are not full time employees and they do not enjoy benefits of many regular employees in different positions. Due to their high-contrast right up in the eyeball themes, it is easier to offer examples rather than abstracted higher layers. Kayfabe sexual orientations and even their marriage on stage which was immediately declared fake is one such example. Another is the bullying of legitimately handicapped wrestlers (which was pulled out of live-TV in other countries soon after).

Left-wing commentators also highlight the other side of the spectrum, often in a somewhat joking way. These involve kayfabe involving the CEO (legitimate!) of the company, where he is on the receiving end. There was a memorable kayfabe 'match' between the said CEO and the later President of the United States, where the CEO had his head shaved off!

On another time, the CEO's limousine exploded on (kayfabe) live TV. Freedom comes unlimited.

Now we have another amusing Gaussian curve!

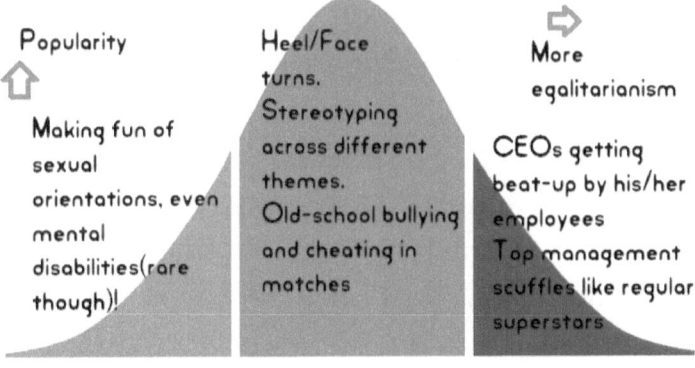

Popularity

Making fun of sexual orientations, even mental disabilities(rare though)!

Heel/Face turns. Stereotyping across different themes. Old-school bullying and cheating in matches

More egalitarianism

CEOs getting beat-up by his/her employees Top management scuffles like regular superstars

Microtransactions

Microtransactions have become one of the most important topics of business in the current generation. With the rising demographics of computer-gaming, the importance of microtransactions was, among others, notable illustrated in a record downvote of a comment by a gaming company regarding the Star Wars Franchise. The importance and the extraordinary effect of these microtransactions are not particularly evident (at least not yet) in the revenue-frontline yet, but are evident in the staggering number of fields it covers. It'll be explained.

Gaming was once regarded as a nerd activity or activities for children. Now, some of the most watched clips in popular streaming platforms are depicting gamers in action. With the rise of mobile gaming, almost every teenager and young adult, even in some of the lesser developed parts of the world, have at one time dabbled with games. Companies were happy, but there was one problem. In many of the developing countries, and newly industrialized ones, people are for the major part stingy with money. They usually do not (simplifying for space) buy games outright, especially the casual mobile gamers. There are also a lot of free games available, so the users often do not even go into the paid games tab. One of the solutions that cropped up was microtransactions, where users have to pay money, usually a rather small amount, to unlock a level which otherwise had to be grinded away (to be played for long duration without much incentive) to complete. This sounds fair, only the hardcore fans would only have to pay a price. The others can enjoy it freely, and the company can still make some money by the way of advertisements.

Microtransactions suddenly gained entry into the scene in a way that was never seen before. Within a few years, it captured all the smartphone and console inhabited parts of the globe. Exceptions were there, however, such as many games from Nintendo, which are usually oriented towards children.

Microtransactions even entered the spheres of applications(as opposed to games), mainly in smartphones. With rapidly rising user-base, also from adults which was previously rather unseen, microtransactions gained enormous powers. Singing collaboration apps, as an example, incorporated microtranscations as a way to gift the singers, both in virtual 'coins' and with real money. Of course, there was a major 'cut' involved.

With great power, comes a great responsibility. Microtransactions were not perfect with its responsibilities, in all its incarnations.

We have to consider the side of the corporations too, and even private individuals who can develop apps and games individually. For one, when the customers were all in Europe and America, the ad corporations paid much more per advertisement shown. This was because the user would be much more likely to buy a product than a teen boy in Bangladesh (incorporating elements of the reach and the median age) currently will do. Secondly, this is applicable more to console games, the cost of development have increased somewhat exponentially with the rise of graphical technologies, which demands more realism from users. What is not so impressive is the tendency to milk the old cow for ages, such as in some popular first-person shooter genres.

The detailed analysis of this particular topic illustrates one very important fact. It is difficult to denote zones in a Gaussian – curve representation, which though in itself is an anathema to purists, is very often used outside courses involving pure statistics. Subjectivity, objectivity, morality, philosophy, all are included in broader scale of applied statistics. It can be a dizzying game of complexity, yet like a breathtakingly beautiful mountain to hike to the summit, even if it was one of many!

Nevertheless of the complexities involved, sometimes in the applied world we need to take somewhat quick decisions. Here is a *pair of* representations, which will not be discussed in more depth to make sure all of the readers are also carefully looking at the curves and not just the text!

A 3D curve (with discrete values in the Z-axis) can be used to denote the changes over time. If they are more difficult to read from, then two 2D representations, like below, can be used.

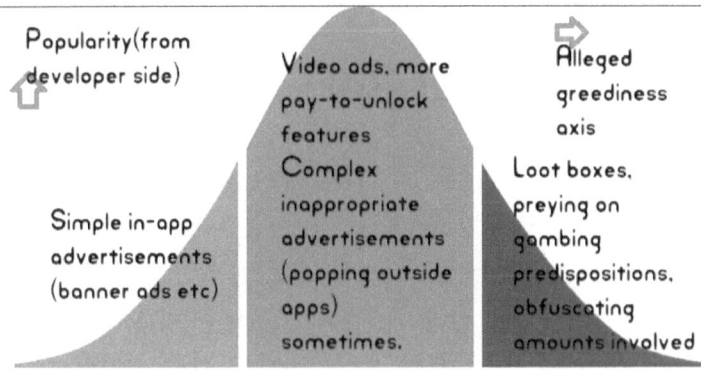

Fig: Microtransactions (including in-app ads in scope)around 2020.

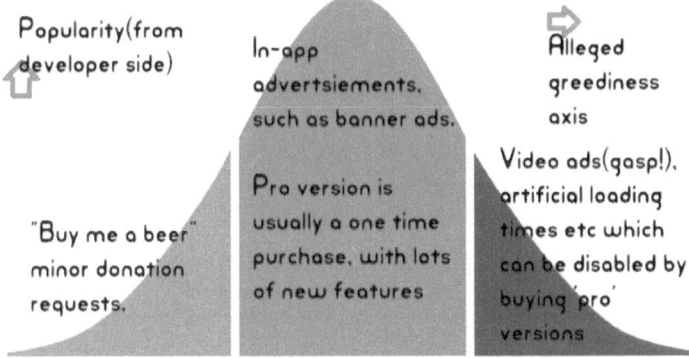

Fig: Microtransactions (including in-app ads in scope)during 2015 and earlier times.

<u>Opportunities</u>

In the majority of the recorded history, humans were living in a society with division of labor. There were leaders, head tribesman, or the kings in one side of the individual wealth axis and the majority was peasants, living on very little in the way of extra income. Kings did have the opportunity to live as peasants for a day, as we learnt in various folk tales, but the reverse was not true. Kings had many opportunities; the bulk of the Gaussian curve did not.

In modern times, however, the majority has so many opportunities to prosper, so much freedom to choose his vocation, that it was perhaps never ever before observed in human history. People living in some remote villages of India, growing up in a very traditional environment, studied to become the CEO of one of the largest companies of the world. It is nowhere specified that things like this are *easy*, because they aren't, or these would not be in the news. But the fact of the current day is that almost anyone can rise to nearby peaks, if not *the* coveted CEO peak, attests to the fact of the opportunity the modern globalized world, in some form of egalitarian and democratic opportunities, pushes the curve from both sides in the number of people vs opportunity cohort.

But no real world is perfect. As evidenced by studies involving affirmative action in the United States, by some conservative researchers, we are very far away from it. Take Thomas Sowell, who himself thought of him as libertarian but was also regarded by others as conservative. In his work, he has shown that even with affirmative action in place, African-Americans and Latinos are still having problems as their classes are too hard in college. If that is the case, these cohorts, in the specific timeframe that Sowell researched, these people are not getting the opportunities of the more productive courses in college than that we would imagine. The situation is somewhat similar in many developing countries, even though explicit objective data is much low and of poorer quality (methodologies, sample size etc).

In many newly industrialized nations, such as India, some trends are being observed which are yet to be quantified extensively but being put forward in hypotheses and minor studies. Many students, who are hailing from rural backgrounds, are trailing far behind than their friends from urbanized areas. There can be many reasons for that, we'll not go into the details, but these trends are pointing to a discrepancy in opportunities which is otherwise present in paper in a casual analysis. Casual and causal should not be mixed.

A good example of a central profession that is very much an opportunity for the majority is the profession of a politician. Of course, few people are cut out for it, but the data shows that politicians hail from very varied backgrounds. That means statistically speaking, when we divide people along the basis of income, parental occupation etc, the dispersion of politicians among them is random(-ish). They can indeed occupy as a portion of the central bulk. Blurring the various factors barring some discrepancies with regard to nepotism and cronyism, these sort of data is something statisticians love! Positions in various government jobs are also sometimes reserved for a specific section of the populace, and these should also go near the centre.

But some people, such as the tribals in various parts of India, are often not as efficiently skilled in picking up office skills which globalization has made plentiful. Even with stimulus in the form of reservations and such, many professions are more or less out of reach for them. Not for *all* of them, it is to be minded, but for the vast majority of them. Obama may have had been the president, but discrimination against the African Americans in still there in the United States, and not in a very minor, statistical anomaly kind of way.

On the other end of the curve, there are still a few opportunities of vocation which only a select few can get. This includes occupations which are possible only by birth, such as belonging to the Royal Family, and ones that require so much specific environmental assistance that only few of the people can realistically acquire. Examples can include founders of some of the largest technological companies of this century. Not only these people had excellent education, very educated parents (who are usually better at finding out latent skills in younger children, besides the genetic assistance) but also had financial support that is rare for the middle class to have, and the added connections in society.

After this slightly longer discussion, we can now put forward a curve!

Acheivable by percentage
of population
⇧

Disadvantaged
due to laocation,
money and/or
societal standings.
The rare ceiling is
often simple
office jobs.

POLITICIAN
(usually around
the mean of the
voting population).
Many simpler
office-jobs(esp.
govt), and most
blue-collar
professions.

⇨Social
prestige
Opportunities tied
to birth, such as
being the King, or
extremely technical
jobs, astronauts,
professors in
Caltech, Fortune-
500 CEOs

Precision

Precision is the measurement of the degree to which repeated experiments concur to the *true value* of the quantity that is being measured. Precision is the backbone of science, because science is based on the reproducibility of hypotheses within a certain degree of precision.

Precision, as many readers will immediately guess, have varied standards in different applications. If we want to measure the density of population in different geographic areas, the precision of our coveted measurement will get lower and lower if we restrict our search in a metropolitan region, then to a city and then to a specific neighborhood of that city. If we are to find the statistical distribution of minor capillaries in a particular person's hand (a bit weird idea for a college project), we might very erroneously check the distribution in another person, where the precision would suddenly take a nosedive.

Therefore to construct a single meaningful curve for this topic, we have to restrict ourselves to what context of precision we are discussing here. Let the context of precision be the precision to which we can measure the length of objects.

Objects can be big, and objects can be small. Objects can be huge, and objects can be tiny. Objects can be ginormous, and they also be miniscule. On the largest of the scale, we have networks of filaments of galaxies, whose specific sizes are very imprecise. Just a few notches below, our own galaxy has a diameter of around 180 kly (Kilo Light year) with a dark matter halo of around 1.9 Mly. This surrounding halo has a precision of +0.4 Mly, which is about 20%! In the largest of the scales, as we see, measurements have a very low amount of precision.

But when we get to more familiar scales, we can see a dramatic rise in the amount of precision that can be achieved. The circumference of the Earth, for example, is expressed up to meters (of course, when discussing specifically about the maximum or the minimum circumference) by the World Geodetic System. This is an enormous achievement in terms of precision.

In even smaller scales, almost anything can be probed up to their nanometer scales, finding their exact length, breadth and height. Of course, these will change temporally based on other environmental variables, but we do have the means to achieve the desired precision.

But anything in the atomic scale and smaller, things start to get murky. Unlike what we read in elementary school books, electrons are not like tiny balls spinning around a bigger nucleus, but they occupy space like a cloud which cannot be resolved unlike rain-bearing clouds. Here the Heisenberg's uncertainty principle gets pertinent. The more precisely we try to determine the particle's location, the less precise we get its momentum. This is very rooted in the fact that it is not a limitation of the current technology, but is the nature of the quantum world itself. Is quantum mechanics absolutely true and there is no better theory to improve our precision in these scales? Science does not deal with absolutes. But currently we do not have a better, widely accepted theory which also can give very precise results the quantum theory can give us, even though it may sound reverse to the imprecision it can inherently carry.

We now have a curve!

Precision achievable

Everyday objects
visible to the eye

Size of
object

At limits of the light
microscope (color
lost), some
challenging electron
microscopy,
subatomic scales
(far-left)

and magnifying
glasses,
sizes of large
geographical
areas,
almost upto the
size of the Earth

Anything to do
with light-years
and beyond,
galaxies and their
structure, groups
and filaments of
galaxies.

<u>Reality</u>

Reality is both subjective and objective. The diameter of the Earth, which was recorded first time in history (note that this does not include prehistory, the time from which written records are not available) by Eratosthenes of Cyrene in 240 B.C., is a figure which almost everyone universally accepts. Few that do not, possibly flat-earthers, hollow-earthers and other believers in conspiracy theories, can be regarded as an anomaly. These figures and related data (another one can be the speed of light), can be regarded as aspects of reality which we can put in a subjectivity and objectivity scale. On the other end we can have the idea of the supernatural, which is very subjective as different people describe them differently, even with similar bases that are dictated from the holy books of various religions.

Perhaps a more interesting version of this curve can be made with a different version of the subjectivity-objectivity scale.

Consider a very common gesture, illustrated as an emoji in recent times, the OK gesture. It is done by connecting the index finger and the thumb in a circle and extending the other fingers. This might be taken, in the first glance, to be a binary value of 1, signifying the sure confidence of the sign being OK, literally.

But that will be, when taken in the more holistic context, something which statisticians refer to as a sampling bias. The vast majority of the readers of this book will be from English-speaking countries, either as a first or a second language, so almost all readers will find this perplexing. But the truth of the matter is the gesture has many different meanings across cultures. We have the meaning of disapproval, which is the complete opposite of OK, and also far more negative ones, which can have unforeseen consequences. Even if we try to incorporate subjects from different ethnicities, we can run into problems of new-culture internalization. American speakers spending a lot of time in Britain can start speaking water as who-tah, and these internalizations can happen subconsciously. Therefore, many of our hand-gestures and other elements of daily interaction have a significant chunk of subjectivity being blended with their objectivity. The objectivity of hand gestures, specifically, is often greater and denser than written or verbal words, as evidenced by hand signs by the traffic police as opposed to the aforesaid signals. Therefore, these are also not miniscule in their objectivity. These interactions can occupy the central bulk because they affect our day-to-day social interactions to a very large degree.

Finding more subjective reality isn't hard, as weed legal states can attest to. Hallucinations induced by other drugs, such as LSD, are also well known. These differ from people to people, as in various kaleidoscopes. There are also other examples in which supernatural beings are interpreted more vividly, but from brain studies, we can more objectively determine the subjectivity in the cases of these drug induced hallucinations.

Finding more objective reality is slighter more convoluted, but not difficult. There are certain reflex actions(related to the human nervous system) which almost everyone shows universally. One example would be the patellar reflex. In this reflex related to the nervous system, specifically some L segments of the spinal cord, if the patellar tendon is tapped with a tiny hammer, the knee extends without any conscious effort by the subject. This is a very good candidate of objective reality. Of course, the basic assumption of "We all live in a matrix" has to be abandoned first.

Another example of a more objective reality is to try out this fun experiment which the author accidentally ran on to. The subject has to stay in a room with a red lamp turned on. For the aforesaid case, the lamp was a multicolor fancy LED lamp whose color could be adjusted by Bluetooth. After 5-10 minutes, the subject can leave the room and gaze at the night sky. Curiously, the whole sky will be tinted in blue. This probably happens as only one of the 3 types of cone cells of the eye, responsible for color vision, was activated. The brain in all probability tried to accommodate for the strange source of singular pure red light which is very rare in nature.

Therefore we can now have a workable curve on the topic of Reality, with the second axis in the subjective-objective scale!

Commonality

Objectivity

Day-o-day interactions, customs of languages, universal hand gestures, accepted pronunciations.

Realities perceived in an intoxicated state, as in under the influence of psychoactive substances.

Nervous system reflexes and perceptions which are independent of cultural upbringing.

Soldier Training

Wars, and in smaller scale, battles, were always there in the history of mankind. How can it be said with such a high degree of confidence? It can be said using the tools of evolutionary biology.

The animals closes to humans in their total genetic similarity is the chimpanzee. But earlier humans were not chimpanzees themselves but very likely a close hybrid (not in the technical sense) between the chimpanzees and the modern man. This is often referred to as the last common ancestor, in the context of evolutionary biology.

By studying the group behavior of the chimpanzees and related apes, we do see group dynamics and a kind of primitive soldier training. On the context of humans, soldier training therefore can be a useful topic to study with the curve.

Modern soldiers have a lot in common with each other across nations and other such groups. The second axis can be used to represent the sophistication of various types of soldier training. The starting point, like the various other topics in this book, can be from the central bulk. Here, the bulk will be the types of soldier training which is commonplace across the world.

The most basic aspects of soldier training, perhaps differing from the common view, are those involving daily discipline; in the form of daily drills and equipment maintenance. This is common to almost all of the formal militaries. Next comes the aspect of physical endurance and of course the use of firearms. Well known international gestures of surrender, road signals and codes are also a part of the central bulk. The use of military communication, in encrypted radio codes, is also very universal nowadays, and can be included near the centre of the curve.

What can be on the region of less sophistication? If we compare the dominant militaries over the centuries, one trend is very boldly visible. Before the advent of firearms, most armies used to wear their uniform with bright colors and marched with flamboyance. Though large -scale marches are still relevant today, it only happens when the battle is over or when it is merely the threat of a battle but not the brattle itself. Here the concept of camouflage, of everything, including the soldiers themselves, is an integral part of soldier training. Entire military books have been devoted to the correct training of camouflage for different environments and the time of the day. Camouflage can be considered as a rather advanced form of soldier training, even though various ancient empires and even animals also use them. The blurriness here can be resolved if we demarcate the types of camouflage. Camouflage during the night using advanced tools of night vision, for example, are a type of soldier training only few select nations can provide. This will be, therefore, on one of the ends of the curve.

Training military personnel on warfare using drones and other autonomous vehicles is a very new field. While it is being dubious, there is no question on its extreme sophistication.

Various indigenous groups around the world, who are sometimes of tribal origin and not modern in the common sense of the word, often use much less sophisticated means of soldier training. This may include yelling, usage of primitive weapons such as arrows and spears and suchlike. Even though they might be sophisticated enough for those individual's own territory (such as against animals and other rival groups), their tools are obviously much less sophisticated in the grander scheme of things.

Now, we have what we were trying to construct.

Popularity

⇨ Sophistication

Use of primitive bladed weapons as primary, flamboyance, lack of modern camoflauge.

Discipline, in drills and maintenance.

Universal military gestures, basic firearms training and military communications.

Use of night-vision equipment, drones, digital counterintelligence, autonomous warfare etc.

Vegetables

Vegetables, when plotted against our age on the basis of preference, show an interesting trend. After we gain the capacity to form long-term memories, which is around the age of 2-4, most of the children show a significant disdain for vegetables. Preferring candies, chocolates or other kiddie treats, the problem is further exacerbated by the fact that their options also stay limited. Which responsible parent, in their right mind, would allow the abundance of cavity-inducing sugary abomination in the diet?

Of course, as our ages increase, most of us do start to include vegetables as a minor preference in our diet, perhaps with some initial begrudging forethought. Then our metabolism slows down as we leave our teens, and the vegetables more or less become a necessary preference. The peak(maximum) however, settles in after a while.

The industrial revolution and the succeeding years started this on a mass scale, and the information age surely nailed it down, when we question about food security. Granted, there is still hunger in this world, but for the first time in history, the overweight-people percentage has become bigger than people living in hunger. This short background is necessary to explain the peak in the last paragraph.

In the age of the done-to-the-bone penetration of social media, looking good has become more or less a necessity for the majority of the people. Even with going to the gym and consuming fat burning pills (which do not work), people without the eternal propensity to be thin do realize that they need to eat more healthily, and in lesser quantities. Vegetables enter the fray here. This causes the peaking of the quantity of vegetables consumed during the middle years. Midlife crisis also enters to the mix for many, and many-a-vegetables get consumed.

The consumption of vegetables does decrease later, but mostly because the overall intake of food decreases as we grow to our senior years. The curve here does not descend here as it ascended during the teens and beyond. It does not show much of a Gaussian slope in the larger X-axis, so we should try another alternative.

Potatoes, as evidenced in one very multicultural international user-generated-content site, is sometimes regarded as a 'God-tier' vegetable. It is very hard to find people who hate potatoes (not of the couch variety though, that is easy to find!). Potatoes are an excellent source of carbohydrates, and can be boiled, fried, mashed, grilled, put inside another food, and can even be burnt to consume the inside. Many other foods also incorporate potatoes in them to make them more delectable, such as the Indian chapattis. Their god-tier is justified, after all!

Another more serious example of the importance of potato lies in Ireland, about 170 years ago. Ireland was very dependent on its local production of potatoes as a staple diet, and during that time a ravaging famine occurred. This resulted in many deaths and large scale emigration, where about 1/5th of the entire population was lost from the island. The fact that people of Irish ancestry is among the topmost in the USA is the result of the famine which was due to the shortage of potatoes.

Corn, carrots and cucumbers are also very popular vegetables, though probably not at all God-tier. Still, they can be put in the central bulk.

There are some unpopular vegetables which are also bad for the ethos of the vegetables. Eggplants and Radishes are some examples, which the author personally found to be repugnant. Although enjoyed by some and is a part of some popular dishes like Tacos, these vegetables aren't that healthy. Radishes, for example, have low nutritional value and they can cause abdominal gas and the ensuing flatulence. Eggplants also do not have that much of nutrition, and can soak up fats and unhealthy oil from surrounding cooking pot, and can go radically opposite to what vegetables should offer, like a conman!

On the other end of the spectrum, we have vegetables which are not that popular due to their debatable taste, but are very nutritious in the right way vegetables should be. The right way is important here, as due to the previously highlighted pandemic of excess-weight (which brings a host of other health problems most readers are very familiar with by now), nutritious no longer means what which has the most calories. Instead, effects such as anti-tumorigenic, anti-oxidative and enrichment by minerals and vitamins is what makes something nutritious.

Brussels sprouts are a prime example of the last kind of vegetables. It is not very popular with kids, with one describing them as "bad smelling, bad tasting and (downright!) evil", but they have useful antioxidants which can prevent cellular DNA damage, can protect ourselves against free radical damage and also has a plethora of vitamins. Broccoli is another example, which according to a scientific paper by PubMed, can protect the heart from oxidative damage. Broccoli also is rich in useful minerals and a few important vitamins. Asparagus can also be thrown in this mix.

Finally we have the curve for vegetables!

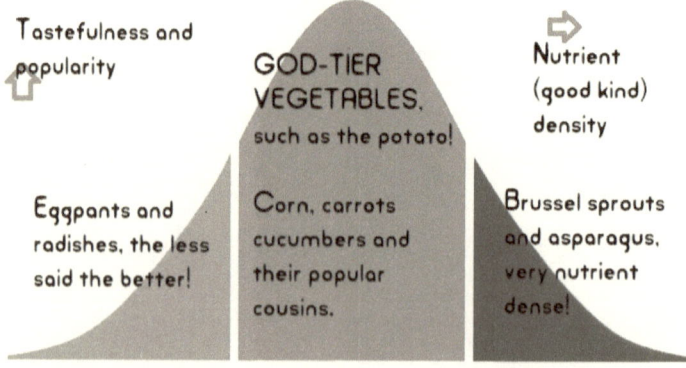

Zodiac predictions

The human mind is an amazing work of art. Fulfill its basic needs of air, water, food and companionship; it is ready to explore new real and imaginary grounds. Ordinary men (and women, as usual) dwell in the past, being nurtured by nostalgia or pained by its pangs. Men who are remembered are often heroes; and they often had exceptional capabilities of dealing with the present during any times of crisis. A third category of men spend a lot of time in the future, trying to predict and morph it. While it may sound a high-tier activity, considerable doubts arise when zodiac predictions are involved in this regard.

Humor aside, it is true that zodiac predictions are still a booming business worldwide. While some people stop at reading the horoscopes in their local magazines, others employ precious gemstones in their rings to change their (predicted and usually negative) future. Sometimes, if the seller of zodiac-related cures are especially crafty, and the recipient particularly wealthy, there are even costlier methods such as those involving voodoo or the Indian yagnas. Suffice to say, these methods do not work, despite their endless promise.

But how can we put them in a curve? We have to consider some history to find one of the ends of it, to which many readers might be a bit unfamiliar. To go by the numbers, we can take a look at India, where it is still an important business, and a lot of history in this regard can be found in the last 100 years itself.

Ancient India, despite some claims by charlatans who say knew the 9 planets (including Pluto), did not mean the planets (*graha*) the same way we do now (although the word is often regarded as a synonym by the natives). Including the five planets up to Saturn, their planets included the moon and the various phases of it separately. One of these phases is often implicated as a zodiacal imperfection (in a wider sense) which must be cured. Curses by the planet Saturn is also regarded as particularly ravaging.

Due to their extra dose of goofiness, and the fact that close-ups of the planets and moons have reached up to the eyeballs of the astrology believers, these have more or less lost their popularity. "Saturn is tilted a few degrees more from normal, and therefore you'll have a bad phase", statements like these are very rare in public astrology talks nowadays. Using the specificity as our item in one of the axes, these can be put these at one end of the curve. Less specific and too-broad(and secondarily outdated, but we'll not be plotting that parameter in the curve).

The middle of the curve will cover the garden variety, very-popular and more precise terminology-use of zodiac predictions. Our current technology can only see starts and the constellations as point sources of light (the fuzziness is due to the Earth's atmosphere, and some exceptions are there, such as the star Betelgeuse), and perhaps this contributes to keeping their mysteries intact. These constellations (in some convoluted way) supposedly predict our future. "People born under this zodiac will be having ill health for the week", "Be careful of strangers in the weekdays" or "There is a chance of large monetary gains this week". These are very common even today in the local language newspapers of many countries, including India.

The more specific predictions get, more people seem to have faith in it, but there is a limit. The author was involved in writings some software for a fun future prediction app once, and there he realized that the specificity should be moderated after a certain extent. Specific timings for events, for example, should be avoided as they would always fail. Too many details also makes the user miss on the "guess the details" aspect. Of course, the app was written for fun and it contained a disclaimer that "there are indeed only a few things where you can waste time more"!

Finally we have the final curve! Thanks!

Popularity with astrology believers

Predictions not observing any scientific hierarchy , e.g. considering moon as a planet. Too vague to raise fear.

Predictions related to the proper astronomical constellations .

Usually vague, but only as much as to get away with!

Amount of detail in prediction

Too precise predictions that is falsified almost everytime. Bad for horoscope, even if in a digital (*gasp*) application!